DISCOUNT HOMEBUILDING

DISCOUNT HOMEBUILDING

How to Save Thousands of Dollars when Building Your Own Home

by Ray Tassin

DRAKE PUBLISHERS INC.
NEW YORK, NEW YORK 10016

Published in 1974 by
Drake Publishers Inc.
381 Park Ave. South
New York, New York 10016

Library of Contress Cataloging in Publication Data
Tassin, Ray.
    Discount homebuilding; how to save thousands of
dollars when building your own home.

    1. House construction—Contracts and specifications.
I. Title.
TH4815.5.T37   692'.8   74-6075
ISBN 0-87749-673-0

Printed in the United States of America.

Contents

Introduction

Many young couples buying their first home acquire a smaller one than they would like, with the expectation of building a larger one later when they can afford it. This was the reasoning my wife and I followed, but when we were ready for the larger house we discovered that building costs had doubled and we still could not afford the home we wanted.

We had a set of plans drawn up by a professional home designer and requested a bid from one of the top local contractors. He quoted $55,000, not counting the land, which we already owned. This was out of the question for a university professor and his kindergarten-teaching wife.

Then I learned that a fellow faculty member had acted as contractor on half a dozen homes, including his own, and had saved up to 25 percent of the total cost. He looked at our plans and estimated the building cost at about $37,000, including a fee of $1,500 to him for helping us, and excluding the cost of driveways and retaining walls, which later ran to about $2,000. Even if the estimate were off by 10 percent, the total cost should not exceed $43,000. With the strong likelihood of such savings, we decided to take the gamble.

This book is a step-by-step account of our experiences in acting as overall contractor for ourselves and, with the assistance of my colleague, employing subcontractors for each stage of the building process and supervising their work.

DISCOUNT HOMEBUILDING

chapter 1
Preconstruction

In building a home, the professional contractor has the obvious advantages of knowing the field, knowing the best suppliers for the materials needed, and knowing numerous subcontractors who are honest, painstaking craftsmen. The professional contractor undertakes, for an agreed-upon fee, to build the house from foundation to interior trim, hiring subcontractors such as electrical, plumbing, concrete layer, and so forth, as they are needed.

When I undertook to act as my own contractor, it soon became obvious that the key to success was hiring and coordinating the work of twenty-five or more subcontractors. This mode of operation requires the builder to do the following:

(1) *Select the right subcontractors.* The best way to get the best subcontractors available is to ask for recommendations from supply firms, particularly the lumberyards, which have field men who act as salesmen solely to contractors. If most of the lumber is purchased from one yard, the field man will be glad to devote a lot of time to helping the builder find good subcontractors.

The builder also can visit new homes under construction to see which subcontractors do good work. This is the way we hired our bricklaying crew. The builder also can get recommendations from other subcontractors.

The professional contractor has a tremendous advantage in that he already knows a number of qualified subcontractors for each step of the construction. Many of

1

them even have their own exclusive crews. The person building his own home has to hunt out the subcontractors and negotiate with them, and some of them will not be as competent as the builder would like.

Another problem that arises occasionally is getting the subcontractor on the job at the time he is needed. In our area, the agreements we reached with subcontractors for specific jobs guaranteed the price for the job but had no clause guaranteeing the appearance of the crew at the stage in construction when their work was required. If at all possible, agreements with subcontractors should include such a clause, as it will avoid costly and patience-taxing delays.

Most subcontractors, particularly electricians and plumbers, are required to have city permits to work in a particular community. The city building inspector checks their work to determine if it meets city standards. This affords considerable protection to novice builders.

(2) *Determine the going rate and negotiate an equitable price with each subcontractor.* The lumberyard field man is helpful here also. In many instances the rate is determined by the square footage of the house. The home designer based his rate on the square footage of the house including the brick. Rates based on the frame only, without the brick, were used by such subcontractors as the rough-in carpenter, trim carpenter, and painter (finish work). Other rates are by the piece, such as $75 per thousand bricks laid. The field man will advise the builder on the methods used in determining going rates for different steps. The rates also are listed with each step in this book.

Bids are made after the prospective subcontractor examines the house plans and specifications. My faculty colleague helped me to negotiate lower rates in several instances.

It also is important to know exactly what the subcontractor's bid includes—what labor he will perform and what materials he will provide. Nothing should be left unspecified when a price agreement is reached.

(3) *Call in subcontractors when they are needed.* Timing is an important factor in dealing with subcontractors. Bids should be negotiated well ahead of time, and estimates given on when each job will be needed. When the time is close, the subcontractors should be notified of the exact date when they can start. Highly skilled builders keep the subcontractors coming one right after the other, avoiding idle days except for bad weather. The novice builder will not be this skilled, but he must keep each successive subcontractor coming in as nearly when needed as possible.

(4) *Check completeness and quality of the jobs.* The builder must make sure each subcontractor does all the work he agreed to do, with the proper skill and craftsmanship. He can't afford to be shy about insisting that all jobs be of professional quality. The builder should keep one set of blueprints for himself and see that they are followed exactly. He also must remember to pick up the blueprints from each subcontractor when he finishes his job, to be used by other subcontractors later.

(5) *Check the purchase of materials.* Each subcontractor is authorized to order the materials he needs and charge them to the account of the builder. But the builder must make certain that all materials charged to him are delivered to his job only, and that all surplus materials are returned for credit. We didn't have this problem, but it is not uncommon. Our suppliers and subcontractors were completely honest with materials. In a few instances, the wrong items were delivered, but they were exchanged by the supplier without question. Poor-quality materials also were exchanged a time or two.

3

The exact number of steps and subcontractors required to build a home varies in different parts of the country, because of variations in climate, tradition, and building codes. Methods, materials, and designs often differ radically.

For example, in some parts of the country houses are almost always built with cellars. Where this is true, cellar construction would be an early series of steps involving at least one additional subcontractor. Often in Oklahoma the cellar is built apart from the house, because of the danger of a tornado's leveling the house and covering the cellar entrance.

In other parts of the country attic rooms are common. We designed our house so that an attic room could be added later with a minimum of additional expense.

Storm windows are becoming common in both hot and cold climates, but these can be added a year or two after the house is built in all but the coldest areas, where they are an initial necessity.

It took a lot of patience waiting for successive subcontractors to come do their jobs. In a few instances iron will was needed to get the job done exactly the way we wanted it. But after looking at hundreds of homes built by full-time contractors, I believe we have a better-built home than most in many respects. The total cost of all items on which a professional contractor had based his bid was $42,185. That was $13,000 under his bid, or a saving of 25 percent.

Any reasonably intelligent and careful person should be able to achieve equivalent savings. There are pitfalls, some expensive, but none that cannot be overcome.

The following are the specific steps we undertook in building our home. Appendix 1 lists the individual subcontractors we employed along the way, and Appendix 2 is a table of steps, subcontractors, and costs for labor and materials. It should be borne in mind throughout the book that the costs refer to our experience only, and may vary widely in other situations and at other times.

STEP NO. 1

SELECT A BUILDING SITE. Soaring land costs today make this an expensive step, but it is a simple one if you buy in a new development area where utilities and paving are part of the package deal. The lot can be purchased with monthly payments and thus paid for entirely before building is begun. Since 1970 land prices have risen each year, and all indications are that these rises will continue.

When we first decided that we would eventually want a larger house, we found a beautiful site—nearly half an acre, with many large trees—near my office. The price was low, because the site was on an unpaved street with

The home site, facing southwest. The post near the center is the northeast corner of the lot (Step No. 1).

four vacant lots and no houses. The opposite side of the street was a small park. By the time we were ready to build four years later, the value of the site had increased by more than 60 percent.

I thought I took all the necessary precautions by having a lawyer check the title and building covenants of the addition, and this is important to do. But I failed to check for utilities. The four-site street had electricity, sewage, and telephone service, but no natural-gas or water lines. For a while it looked as though we might have to pay the high cost of these lines, but the gas company eventually installed its line without cost to us, and the city brought the water mains to the corner of the block. All these points should be checked *before* buying the land.

STEP NO. 2

SELECT A HOUSE PLAN. Four sets of the house plan are needed. Stock plans are available from dozens of magazines and most local home designers. Prices range from $20 to $500.

However, we wanted custom plans based on a floor layout we had spent three years developing. An Oklahoma City home designer—an architectural draftsman, not an architect—designed it from our drawings for 12¢ a square foot, a total of $342. The price seemed high, but it would have been foolish to build an expensive home without getting exactly what we wanted. This price, like all the other costs cited in this description of our home-building experience, was based on many factors, including the state of the economy at the time and the region in which we were building. In order to provide some basis for comparison, however, I shall continue to cite our actual costs, urging the reader to keep in mind the possibility of wide fluctuations under differing circumstances. It should be noted also that many townships and municipalities require the seal of a licensed architect on blueprints.

The house plans must conform to local building codes, and the designer will take care of that. But the plans also must be adapted to the lot. Flat lots are the easiest to work with, but often lots are sloping, and this must be considered in designing the plan. Our lot had more slope from east to west than we had realized, and thus required more earth moving than expected, but it still worked out with installation of a retaining wall along the west property line.

STEP NO. 3

SECURE FINANCING. When the house plans were completed, we took a set to the finance company that held a mortgage on our current house. We had never been late with a payment in ten years, so we thought we should have an excellent credit rating with the firm, and we did. In two weeks they notified us that they would lend us the $35,000 we asked for *just as soon as the house was completed.* That meant we would have had to have the cash to pay for construction, in which case we wouldn't need a loan.

So we had to go to a lending agency that made construction loans. That took another two weeks, a delay that could have been avoided if we had known the difference between a mortgage and a construction loan.

The terms of the loan we negotiated specified that we would receive $12,500 when the roof was completed, another $10,000 when the brick and sheetrock were added, and the balance when the house was completed. This meant we needed several thousand dollars to carry us until the first payment from the lending agency, but we had planned to put at least that much into the house, so that was no problem.

STEP NO. 4

SECURE BUILDING PERMIT. Next we took a set of house plans to the local building inspector and filed an application for a building permit. He checked the plans and made certain they complied with all local ordinances and the building code. This check, and the building permit, cost $336. An additional $60 was added for tapping into the city water line, making nearly $400 before we got out of city hall ready to start building.

If we had chosen to build outside the city limits (where we could have avoided these costs, but also would have been without police or fire protection), we would have had to install a water well and a septic tank, the cost of which ranges from several hundred to several thousand dollars, depending upon the location.

STEP NO. 5

PURCHASE BUILDER'S RISK INSURANCE. Builder's risk insurance covers all workmen while on the job and insures against theft, vandalism, and other losses during construction. I found that the firm from which I had a policy on the house I already owned had a unique combination plan: a builder's risk policy during construction that became a homeowner's policy when we moved in.

STEP NO. 6

OPEN CHARGE ACCOUNTS. Contractors receive discounts on a major portion of the materials that go into the house—up to 50 percent on lighting fixtures and lesser percentages on carpeting, lumber, and other items.

Any person building a home in a sense becomes a "builder" or "contractor" and can get these discounts by checking around and by so identifying himself when he opens charge accounts with suppliers. The yellow pages of the telephone directory are one source of finding suppliers in your area. Another way is to question subcontractors at construction sites; an inquiry to each firm recommended will disclose which ones grant discounts. Some suppliers charge the same price to all purchasers.

The major materials expense will be for lumber. In our metropolitan area only three lumberyards sold primarily to builders and offered discounts. You can learn which ones in your area do so by (1) calling them on the telephone, (2) driving around housing-development areas to see which yards are delivering the bulk of the materials, and (3) asking other builders.

When you have made your selections, you then open accounts with them, in order to get materials without having to pay for each load as it is delivered, and to get the discounts.

You also need to open accounts with a ready-mix concrete firm, a brick company, a builder's supply company (for hardware and windows), a carpet company, and so on.

When accounts have been opened, each subcontractor is notified—as he starts the job—where to buy the materials he needs, using your charge account. Most subcontractors are paid for labor only; the builder pays for materials separately, with some exceptions, such as the tile man.

chapter 2

The Base

STEP NO. 7

SITE PREPARATION. Actual construction starts with preparing the "pad" on which the house will stand. This means having a tractor or bulldozer level the land. If our land had been fairly flat already, without rock, a tractor could have done the job in two hours at $9 an hour. But our house site had solid rock, so we had to use a bulldozer for three hours at $15 an hour, for a total cost of $45. Remember that this price is what we paid in our area, at the time we built, and is used for purposes of comparison only.

STEP NO. 8

STAKING AND MARKING THE SITE for the house involves driving stakes in the ground and marking lines for the exact place where the foundation will be built. In our case lines were marked on the ground with powdered lime, using the stakes as a guide. The lines have to be exact and perfectly straight. This work was done by the concrete subcontractor who also dug and poured the footing and stemwall (see Steps No. 9 to 12).

It is important to know local ordinances on site marking. Our city required the house to be at least 25 feet from the front property line and 5 percent (of 110 feet) from the side property lines.

We built the house 50 feet from the front property

White lime is used to mark the exact place where the footing
will be dug. This view is facing west by northwest
(Step No. 8).

Another view of the site, facing north, with lime marks
visible. In the center background are the buildings of
Central State University. A small park separates the home
site from the campus (Step No. 8)

line and left 7 feet on the east side and 14 feet on the west side. It was placed off center because the west side was lower than the east, and centering the house would have required considerably more earth fill on the low side. Even so, a 2-foot retaining wall had to be constructed on the west side.

STEP NO. 9

DIGGING THE FOOTING came next. This usually is done by a trenching machine, but solid sandstone on our site made it necessary to use a backhoe. In this case, the entire east end and about 20 feet of the front and back trenches touching the east end resisted the backhoe. This made it necessary to bring in a jackhammer for two days at a cost, in our case, of $60 for jackhammer and $72.31 extra for labor. The backhoe charge was $65, making a total cost of $197.31 for digging the footing. But a house built on rock sets a lot more solidly than one built on earth.

A jackhammer is used to dig the footing in solid rock (Step No. 9).

Still more rock requiring the jackhammer (Step No. 9).

A backhoe dug the footing where the rock was not a
problem (Step No. 9).

13

STEP NO. 10

PUT IN STRUCTURAL STEEL. A concrete footing gets its strength from the steel rods inside it. Our building code required two rows of ⅜-inch rods with 20 percent overlap at each joint of each row.

Also required at this point was an inspection by the city building inspector to determine if the trench for the footing was the proper depth—a minimum of 18 inches—and if the minimum amount of steel was used. The sub-

The first concrete is poured into the footing trench. The steel rod in the center foreground was used to strengthen the footing (Step No. 11).

contractor requests this inspection when he is ready for it, but the builder must make certain the inspection is made before the concrete is poured.

STEP NO. 11

POUR THE FOOTING. Concrete for the footing is purchased from a ready-mix concrete firm and poured directly from the mixer truck into the trench. It is necessary for this footing to be smoothed out fairly evenly so that the stemwall will set properly on top of it. For our house it took 34 "yards" of concrete at $16.70 per yard, or $597.16, less $8.50 cash discount, leaving a net cost of $570.66 for materials.

STEP NO. 12

SET FORMS, POUR STEMWALL. The concrete crew next sets up the forms for the stemwall on top of

Wooden forms are set up for the stemwall that sets on top of the footing (Step No. 12).

the footing. Another 7.25 "yards" of concrete was then poured into this from ready-mix concrete trucks at a cost of $129.56. Steel band irons to hold the forms together at the bottom, and anchor bolts placed in the top of the stemwall to which the walls are later attached, cost $24.00. We used the same concrete subcontractor for Steps No. 11 and 12, but often separate subcontractors are used.

Another view of the setting up of forms for the stemwall
(Step No. 12).

16

Concrete is poured for the stemwall (Step No. 12).

The finished stemwall after the forms are removed
(following Step No. 12).

STEP NO. 13

SEWER LINES are the first lines to be installed under the slab, because they need to be the lowest. It is important that all these lines drain gradually toward the line that goes outside the stemwall and to the main line. There must be no low spots, no sharp turns (90 degrees or more), and no poorly sealed joints.

If the line slopes too sharply, liquid will move so rapidly that solids will be left behind in the lines, eventually causing the lines to stop up.

We ran into one additional problem. A sewage "lift station" at a lower point on the main line pumped sewage up hill under pressure past our house. During construction the main line became stopped up beyond our house. Because our house was the lowest on the line between the stoppage and the lift station, several hundred gallons of sewage was pumped into our house. Fortunately, con-

The first sewer line is laid. The sand in the right background is used to lay the lines with the proper slope so that the sewage will flow freely to the main line along the back of the lot (Step No. 13).

struction was not far enough along for any damage to be caused. A valve was installed in the line from our house to the main line, so that backed-up sewage could no longer get into the house.

About the only precaution the homebuilder can take to avoid this sort of thing is to install a back-up valve in the first place. They are not too expensive, although most houses never need them. The plumbing subcontractor takes care of this and Steps No. 15, 25, 47, and 48.

The plumbing subcontractor uses the backhoe to dig a trench for the sewer line to connect with the sewer main (Step No. 13).

A sewer line joint is cemented (Step No. 13).

STEP NO. 14

ELECTRICAL, HEATING, AND AIR-CONDITIONING GROUNDWORK. When the plumber fin-

A network of heating-cooling vents is placed under the floor (Step No. 14).

ishes the sewage lines, the electrical subcontractor installs the ductwork for heating and cooling. Concrete is poured around each joint of the ductwork, and around each vent outlet, to insure that it remains securely attached while fill sand and water lines are put in above the ducts. A single subcontractor, the electrical subcontractor, usually does this job, along with Steps No. 17, 21, 26, and 49, although sometimes separate subcontractors are used.

This method of heating and cooling ductwork is, of course, only one of several types available; it was especially suited to our geographic location and climate.

Concrete is poured around each joint of the heating-cooling ducts to seal them permanently (Step No. 14).

STEP NO. 15

WATER LINES are the third and last of the lines under the floor slab, because they are the lightest and smallest in size. It is vital that they be copper lines, with no joints below the slab. No reputable plumbing subcontractor would violate either of these points, but it is a wise precaution to watch for this anyway—you might have hired the wrong subcontractor.

Plastic drainage pipe is approved for our city, but it is a lower-quality line and its use is not permitted in all areas. The builder needs to make sure he gets the quality called for in his blueprints, and not a substitute grade.

Sewage and water lines together are what the plumber calls the groundwork of his job. Normally he is paid 30 percent of his total bid for plumbing when he has completed this portion (Phase I) of the job.

Under-the-floor water lines are placed on top of the heating-cooling ducts, which were placed on top of the sewer lines (Step No. 15).

STEP NO. 16

FILL SAND BELOW THE SLAB. The concrete finisher orders additional truckloads of fill sand dumped inside the stemwall to fill the space to within 4 inches of the top of the wall. He then spreads it out, tamps it down with a tamp or rolls it, then waters it. The watering is supposed to help pack down the sand, but in our case a 5-inch rain the night after we smoothed out the sand did far more packing and settling than a garden hose— and it delayed construction until the sand dried out. For

Loads of sand cover all the lines that go under the floor
(Step No. 16).

our house it took twelve loads of sand, twelve yards each, to fill the stemwall. At $20.50 a load, the total was $246.

The sand is smoothed out evenly over the lines, filling all the space up to the shoulder about four inches from the top of the stemwall (Step No. 16).

STEP NO. 17

INSTALL TEMPORARY ELECTRICAL OUTLET.
At some point between the beginning of construction and this stage, preferably about now, the builder notifies his electrical subcontractor to install a temporary elec-

trical line and pole near the back of the house. Electric power will be needed by various subcontractors from this point on. There is no charge for the installation—only for the power used.

STEP NO. 18

BUILD FORMS FOR VARIOUS FLOOR LEVELS. If the floor is to have more than one level—and nearly

Forms are placed for the various levels of the floor. Here the worker is standing in what will be the sunken living room. On this side of the board the floor will be seven inches higher (Step No. 18).

all do, at least for the garage—forms must be built for the various levels. We wanted a sunken living room, and a brick floor for part of the house. These required three different levels of slab, plus a fourth level for the garage. The brick section was outlined with 2-by-4's and the living room with 2-by-6's. Then wire mesh was laid on top of the sand.

All of these jobs were done by the concrete finisher, along with Steps No. 16, 20, and 21. The four steps comprise Phase II of the concrete work, concrete finishing.

Forms for the three levels of concrete are complete here. At left is the lowest, the sunken living room. In the center is the long hallway that will eventually be covered with floor brick, so it must be poured lower than the area to the right that will be carpeted (Step No. 18).

STEP NO. 19

TERMITE PROOFING. Spraying of the sand under the slab against termites is required by all lending agencies. A liquid chemical is sprayed onto the sand a few hours before the concrete is poured, supposedly guaranteeing against termites for at least five years. For an area 45 by 89 feet, the cost was $40, with no additional cost for spraying under the porches later (Step No. 43).

Spraying chemicals on the sand to prevent termites (Step No. 19).

STEP NO. 20

PLACE INSULATION AROUND THE STEM-WALL. Just before the slab is poured, pieces of insulation are placed just inside the exterior stemwall to protect the slab against cold weather. The slab is poured on top of the insulation. The concrete subcontractor does this job for the cost of the insulation material, but most builders no longer regard this step as necessary.

STEP NO. 21

POURING THE FLOOR SLAB. The final step for the bottom of the house is pouring the slab and smoothing and leveling it. This also is the end of Phase II of the concrete work.

We used a single subcontractor for all three phases of concrete work, but usually a separate subcontractor is used for each.

Pouring the slab is a critical step. If the weather is too hot the slab might crack. If it is too cold the concrete might freeze. It must be worked out smooth and level if the floor coverings are to look attractive.

One end of our living room came out 2 inches lower than the other, but it wasn't noticeable because the room was 40 feet long. The dining room also was lower in one

The first part of the floor slab is poured (Step No. 21).

28

Another view of pouring the floor slab (Step No. 21).

The concrete for the floor slab is quickly leveled out. Workmen
will then "float" the concrete: that is, jiggle it on top with a
flat metal surface with holes in it. This causes the crushed
rock to go down and the small grains of sand and cement
to rise to the top: it is necessary to properly "seal"
concrete. Concrete not properly sealed will flake off and
crumble at the top (Step No. 21).

corner by 2 inches, but the floor was covered with brick and leveled out at that time. The only cracks were in the garage floor, where they caused no trouble. Some surface cracking is normal on large-area jobs.

Our labor costs on Phase II (Steps No. 19 to 21) came to $431.16, and the materials cost $925.81. That wasn't bad for 2,742 square feet of living space, plus a two-car garage and a 400-square-foot workshop behind the garage. Remember, however, what was said about variations in costs, depending on area and the general state of economic conditions.

Finally the floor slab is smoothed out with hand trowels (Step No. 21).

chapter 3

Walls and Roof

STEP NO. 22

ORDER SPECIAL MATERIALS. About this time, certainly not much later, it is wise to select and order brick, light fixtures, windows, floor tile, bath tile, and any other items that might take time for the supplier to secure. For example, brick might take several months if it is manufactured out of state; or, if nonstandard windows are needed, they must be custom made. We had two problems, obtaining living-room light fixtures we wanted, and having a special floor brick shipped from Ohio to Oklahoma. But we had ordered both well ahead of time, so the problems really were minor.

STEP NO. 23

ROUGH-IN CARPENTRY AND WINDOW INSTALLATION. The first step above floor level is the rough-in carpentry, or framing of the house. This consists of putting up the walls—2-by-4 studs covered by sheathing (plyboard or rockboard) on the outside to brace the walls and help insulate the house—outside door sills, ceiling joists, roof rafters, and windows. It is important that sill insulation be placed under all outside stud walls; no air must get in between the concrete and the sill plate, which is the bottom of the stud wall.

This is the most critical step in all the building process, although others are important. No matter how careful the rest of the subcontractors are, they cannot

The first wall goes up (Step No. 23).

improve on the skeleton of the house. Some tract homes I looked at had walls visibly leaning and ceilings that visibly sagged.

Fortunately, our rough-in carpenter and his crew took considerable pride in their work. They checked each wall, each joist and rafter, and squared and plumbed everything. I doubt if even a majority of rough-in carpenters are this skilled and careful. About all the builder can do is watch them and insist that they correct the more obvious mistakes.

Another wall in place (Step No. 23).

A nail gun is many times faster than the old-fashioned
hammer and nails (Step No. 23).

The carpenters order the material from the supplier as it is needed. Our carpenters were careful to make economical use of lumber, hunting up short scraps when short pieces were needed instead of cutting them from longer pieces. When they had finished, the only scraps were tiny pieces. This saved us hundreds of dollars.

It took four carpenters fifteen work days to do our rough-in. In our case, the charge ran 70¢ a square foot based on the living area of the house, a $75 flat rate for the garage and work area behind the garage, $50 for the 10-by-36-foot porch on the back, $50 for gun nails, and $25 for a change I insisted on in the ceiling joists. The total rough-in labor came to $2,119 and materials to $5,157, plus $420 for windows, which are ordered and installed by the rough-in subcontractor. These prices could be higher or lower in other areas.

The ceiling joists come next (Step No. 23).

Finally the first rafter goes up (Step No. 23).

Now come the supports for the rafters (Step No. 23).

A lattice-type deck is nailed across the rafters to hold the shingles (Step No. 23).

homes, but for some unexplained reason the local lumber-yard sold our roofer 18-inch shingles at a dollar a square less in cost. The longer shingles are thicker on the big (exposed) end, making them more attractive and adding perhaps five years to their life. The materials totaled $2,665.70, nearly $1,000 more than they would have cost had we bought them a short time earlier (but perhaps considerably less than if we were buying them today). This was the largest single increase in cost above the original estimate of the faculty colleague helping us build the house.

Here are the water lines for the automatic washing machine (Step No. 25).

STEP NO. 25

"TOP OUT" PLUMBING. Phase II for the plumber was the "top out" plumbing, including the water lines to and drains from the kitchen and utility-room sinks, two washbasins, one bathtub, two hot-water tanks, one shower, and three outside faucets; and gas lines to two hot-water tanks, the central heating unit, and an outlet in the work-shop area behind the garage. For labor and materials the plumber received $682.46, the second of his three payments.

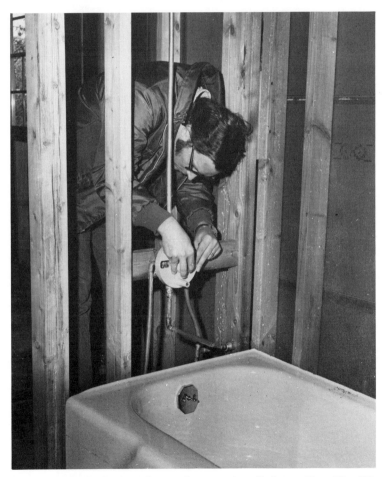

The bathtub and water lines are installed next (Step No. 25).

The final step of the "rough-in" is to install the windows (Step No. 23).

STEP NO. 24

ROOFING. On our house, four roofers spent four days putting on the cedar shingles: some 50 squares at $6.50 per square (regular pitch), and 8.7 squares at $10 per square (steep pitch, requiring scaffolding), a total of $412 for labor.

Ordinarily, 16-inch shingles are used for middle-level

The first shingles are nailed on (Step No. 24).

Shinglers near the top (Step No. 24).

39

STEP NO. 26

ELECTRICAL, HEATING AND AIR CONDITION-ING, Phase II, includes installation of all the electrical wires to the light boxes (but not the outlets or light fixtures yet), the breaker-switch box through which all electrical power moves, and the return air ducts in the attic. We ran into no particular problem here. The electricians followed the blueprint and specifications perfectly, and at our request added a few minor things, such as a television antenna wire to the attic. All subcon-

Electrical wiring is installed (Step No. 26).

Overhead electrical wiring
(Step No. 26).

Boxes for electric switches and
outlets are installed
(Step No. 26).

Return air ducts are installed in the ceiling (Step No. 26).

tractors make an extra charge above the bid price for such additions. Labor and materials totaled $1,427.50 for both Phases I and II (Steps No. 14, 17, and 26).

The return air ducts are wrapped in insulation to keep the air warm or cool (Step No. 26).

The main return air duct eventually stretches 95 feet, counting two turns (Step No. 26).

STEP NO. 27

PRE-WIRING FOR TELEPHONES comes next. If this is done before the walls are covered with sheetrock, the cost is much less. Also, if a home is already completed, the telephone company may require that an electrician run the wire in the attic and down the walls; phone companies no longer do this work in many areas of the country. If the company does the wiring after construction, the lines will be run outside the house, directly through the wall, and around baseboards, unsightly in anyone's new home.

We had five telephone outlets installed during construction, at a cost of only $12.24 for the wiring. The outlets were installed in Step No. 54.

Pre-wiring for one of the five telephone jacks (Step No. 27).

STEP NO. 28

ELECTRICAL INSPECTION. Most cities require inspection of the electrical wiring after installation is completed and *before* insulation—or at least sheetrock—is installed. A stamp of approval is placed inside the breaker-switch box. The builder requests this inspection the day before he wants it. The charge normally is included in the building permit fee.

My faculty colleague who helped me build the house inspects the wiring coming into the breaker-switch box (Step No. 28).

The city building inspector places his official seal in the breaker-switch box to show the wiring has passed inspection (Step No. 28).

STEP NO. 29

INSULATION OF THE WALLS is essential in almost all climates. A subcontractor and small crew can com-

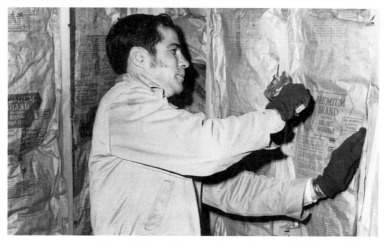

A special stapling gun is used to staple the insulation between the studs (Step No. 29).

plete the job in a couple of days. All wiring and plumbing that go into the walls must be installed first. Despite the large size of our house, this step cost only $208 for both labor and materials.

Scraps of the insulation can be stuffed in cracks around the windows to cut down on future heating and cooling bills, but the builder must do this himself.

Strips of insulation are tacked along the eaves (Step No. 29).

STEP NO. 30

SHEETROCK, TAPE AND PLASTER, or drywalls, normally are installed by a two- or three-man crew. The sheetrock is nailed up first on walls and ceilings. Tape and plaster are put over the joints and nail holes, and the plaster is then sanded smooth. We paid $717.64 for labor and $858.25 for materials for this step.

Plastering over the cracks and nail heads (Step No. 30).

Unloading sheetrock (Step No. 30).

Sheetrock goes on the ceiling first (Step No. 30).

Another view of plastering the ceiling (Step No. 30).

Smoothing out plaster on the cracks and nail heads of the
walls (Step No. 30).

STEP NO. 31

*"BLOW" TEXTURED MATERIAL ON THE CEIL-
ING.* In better homes, a heavily textured off-white material with sparkle is blown on the ceilings instead of
painting them. It lasts about three times as long as the
paint on the walls, and is more attractive. This step may
be done by the drywall crew or the painting subcontractor. For our house, the cost was 5¢ a square foot, a
total of $175.00, for labor and materials. However, merely
painting the ceiling would have involved cost, so the price
of the textured material was partially offset.

"Blowing" the texture on the ceiling (Step No. 31).

"Blowing" the rough material and sparkles on the ceiling (Step No. 31).

STEP NO. 32

INSTALLATION OF THE GARAGE DOOR AND OVERHEAD ATTIC LADDER is one of the easier tasks for the builder. Companies specializing in this work come out to the site, check the openings, offer a wide selection of styles, and then make a bid that includes both the doors and the installation. For our choices, the cost for an oversized two-car garage door was $292.74 installed; $192.61 for the electronic device that operates the garage door by pushing a button in the car or on the garage wall, and $28.51 for the attic ladder installed.

Installing the garage door (Step No. 32).

STEP NO. 33

TEXTURIZING WALLS, AND PAINTING OUT-SIDE TRIM are Phase I of the painting subcontractor's

Part of the outside base coat can be sprayed on (Step No. 33).

job. "Blowing" the textured material on the ceiling would have come before these items if the drywall subcontractor had not already taken care of it. Otherwise, the first task is to texturize the walls; that is, give them a lightly rough surface, if desired. Then the base coat is painted on the outside trim. This should be done at this point because much of the paint is sprayed on and would mar the brick if the brick were already up.

Painting the base coat on the outside trim (Step No. 33).

chapter 4
Bricking

STEP NO. 34

BRICKLAYING is another very critical step, both the inside fireplace and the outside veneer. It is important to hire bricklayers skilled enough to build a fireplace

The first bricks go up on the end of the house (Step No. 34).

that will work properly and outside walls that will be in perfect alignment for each row of bricks. Looking around new housing areas, we saw a lot of houses where the brick did not fit the above description. On our own outside walls we found only one spot where the rows were not in perfect alignment, and a protruding wall may have been at least partly responsible.

The bricker who had worked for my faculty colleague on previous occasions was not available for my house, and the brick plants were unable to recommend anyone else who wasn't booked up for several months. So my helper simply drove through new housing areas looking

Bricking the outside (Step No. 34).

56

Building the fireplace (Step No. 34).

at brick jobs until he found a crew doing high-quality work. He made a deal at $75 per 1,000 bricks, an additional $150 for the fireplace, and $30 extra for scaffolding needed to brick a two-story gable on the front of the house. The total cost of labor was $1,250; the brick, mortar, and angle irons used for the fireplace came to $1,300.87.

Ordinarily, the brick is then washed with acid to clean it of excess mortar. Such a bath would have removed the white frosting from our colonial-looking brick, so we had it wire-brushed instead.

Building the fireplace, second view (Step No. 34).

chapter 5
Interior Finishing

STEP NO. 35

BATH AND FLOOR TILE AND FLOOR BRICK should be installed next, before the trim carpentry work.

Leveling the floor for floor brick. A nearly dry, almost crumbly mortar is used in order to get it perfectly flat and smooth (Step No. 35).

Laying the floor brick, including the side that drops down one step into the sunken living room (Step No. 35).

In some instances the trim work is done first, but not usually. Companies selling tile and brick know the best installers. A single subcontractor can do both tile and floor brick, as was the case with our house, or separate subcontractors may be used. We used the same subcontractor who installed the shower doors, towel racks, and soap dishes that are built in as part of the tile.

We paid the subcontractor $1,095 for labor and tile in three baths, and paid $420 for the brick used on the floor of the entry, the hall, and the utility, kitchen, and dining rooms.

Grouting the floor brick. This is achieved by covering the entire floor with mortar, packing it solidly between the bricks, then removing the surplus mortar from the top by scraping it, washing it, and giving it an acid bath (Step No. 35).

The floor brick gets an acid bath to remove mortar
(Step No. 35).

Covering the shower walls with mortar to which the tile will
be attached (Step No. 35).

Applying the bathtub wall tiles (Step No. 35).

STEP NO. 36

TRIM CARPENTRY (PHASE I) includes all the cabinet work, door and window facings, bookshelves, paneling, hanging of doors and shutters (inside and out), posts, accent pieces, and other visible woodwork. Quality of craftsmanship shows more here and in the next step (painting and staining) than anywhere else.

Lumber is more expensive for this step, also. The trim crew orders the lumber as it is needed. If they are careful

Installing the wainscoting (Step No. 36).

Putting the molding around the top edge of the wainscoting (Step No. 36).

Fitting a small strip of mopboard (Step No. 36).

Installing window sills (Step No. 36).

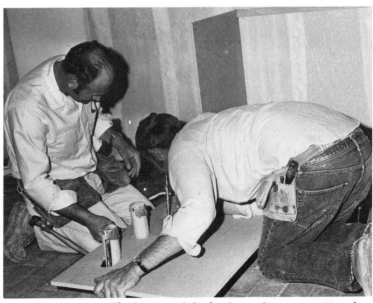

The bottom of the kitchen cabinet goes into place
(Step No. 36).

they can save many hundreds of dollars by using every piece of wood possible, wasting nothing but scraps too small to be used.

The homeowner will need to spend considerable time on the job each day explaining just how each item is to look. The blueprint should contain all of these details, but at least some are bound to be missing. In our case, there were no sideviews of bookshelving in two of the rooms. The day the carpenters started on this part of the job we were late, and they had part of the shelving in place when we arrived. It was not at all what we wanted,

Hanging the front door (Step No. 36).

67

The first vertical wall (bulkhead) of the kitchen cabinets goes up (Step No. 36).

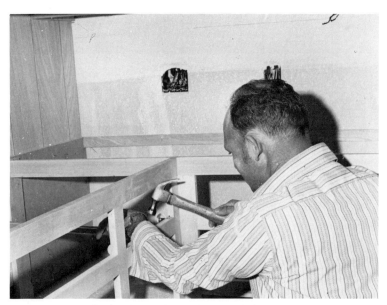

The kitchen cabinets begin to take shape (Step No. 36).

so we had them rip out what had been built and start over. The subcontractor charged us $160 for extra labor, and we ruined at least $100 worth of lumber. Had we been there when they started, or had the plan been complete in every detail, this would have been avoided.

Deviation from the plan involving additional man-hours will result in extra charges also. The subcontractor should get the builder's permission before making a change, and before doing work for which he will charge.

Building the kitchen pantry in the back side of the fireplace (Step No. 36).

A china cabinet also is built into the back side of the fireplace, opening into the dining room (Step No. 36).

Labor for this step was part of the overall trim work, charged for as a single item with Step No. 52. We paid for all materials, even sandpaper, nails, and glue.

The overhead section of the kitchen cabinets goes up (Step No. 36).

Adding shelves over the washer and dryer (Step No. 36).

The mantel is attached to the front of the fireplace (Step No. 36).

Putting up the overhead beams in the living room (Step No. 36).

A cabinet door is hinged into place (Step No. 36).

Building a banister along one side of the living room, separating it from the hallway (Step No. 36).

72

The toprail is added to the banister (Step No. 36).

Sanding the toprail (Step No. 36).

Outside trim work includes installation of shutters
(Step No. 36).

Another shutter goes up (Step No. 36).

STEP NO. 37

FINISH WORK includes the painting of all inside walls and outside trim, and the staining and lacquering of all woodwork. These jobs are all done by a single subcontractor, the trim finisher or painter. He provides all materials, which he can buy cheaper than you.

It is important to choose the stain and paint ahead of time, to avoid delay. Don't rely on 1-inch squares on a paint chart to show the actual color. The supplier will provide samples of a few for final selection.

Spraying stain on the book shelves (Step No. 37).

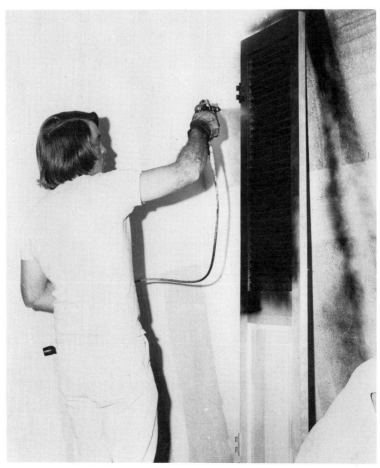

Spraying stain on a half-door (Step No. 37).

Woodwork stain is a bigger problem than paint: paint can be changed, but once the stain is on the woodwork, there is no turning back.

The cost for the finish work came to 60¢ a square foot, or $1,645.20.

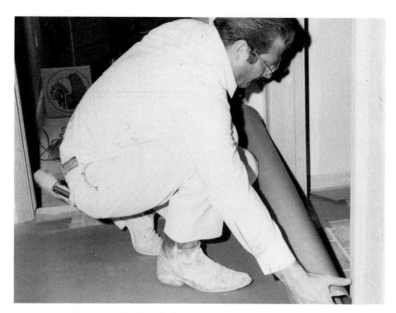

Covering the brick floor to keep paint off of it (Step No. 37).

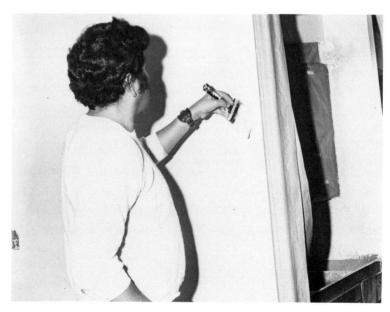

Brush painting the walls close to the woodwork (Step No. 37).

Wiping off the surplus stain (Step No. 37).

Rolling paint on the walls (Step No. 37).

STEP NO. 38

INSTALL CABINET TOPS. Another simple task for the builder is the installation of cabinet tops. Any metropolitan area contains numerous companies that will come to the house, measure the space to be covered, and quote a price for the job, both labor and materials. Several good brands of material are available, including Nevamar and Formica. A few days later the subcontractor will return with the tops and install them.

It is a little less expensive to have the trim carpenter build the cabinet tops on the job, but that means that the edges and backsplash must be squared off. Putting a hump at and rounding the edges keeps water from running on the floor. Rounding the backsplash keeps it all one piece of material and avoids having a crack at the back of the cabinet top. However, this rounding can only be done in a shop, where the material can be heated. That is why the cost is greater, but the difference is less than $100. A full 20 feet of cabinet top in our kitchen and three feet in our utility room totaled only $313.63 installed.

STEP NO. 39

INSTALL VANITY TOPS. Similarly simple for the builder is the installation of vanity tops in the baths. The favorite style today is man-made marble, with the wash-basin top made all in one piece. The result is beautiful beyond all other materials, and cheaper than any, even the laminated plastic used for kitchen-cabinet tops. The subcontracting process is the same as for the cabinet tops. A marble company sends a representative, who measures and quotes a price. The tops are then manufactured in the firm's plant, brought out to the job, and installed, all for a single price. We used three 5-foot vanity tops with washbasins built in, a 5-footer without washbasin, all with 4-inch backsplashes, and a 3-foot-square table top in the kitchen, all for $305.62 installed.

chapter 6
Outside Finish Work

STEP NO. 40

CLEANUP, INSIDE AND OUT. The various sub-contractors tend to leave their mess behind. By this stage there is quite an accumulation of litter, both inside and out. We hired a high-school senior with a pickup truck at $2 an hour for his labor, $4 a day for his pickup, and 50¢ per load dump fee. He worked several times during the construction, for a total cost to us of $125.00.

STEP NO. 41

GRADING for driveways, walks, and porches is done with a heavy-duty tractor. Sand is hauled in for the base, because it compacts better than regular earth, thus reducing the likelihood of later cracking. Tractor work cost us $9 per hour for three hours. Sand and fill dirt totaled $166.74.

Grading for the driveway (Step No. 41).

STEP NO. 42

BUILDING FORMS FOR DRIVEWAYS, WALKS, AND PORCHES is Phase III of concrete work. I had carefully cleaned and salvaged the form lumber from earlier concrete work, so we needed only $76.35 worth of additional materials. Labor was figured in with Step No. 44.

One problem we ran into was drainage from a sidewalk behind the house. The walk was built 4 inches lower on the end next to the porch and funneled water from the roof right onto the porch. I had to erect a barrier to prevent this. Driveways and walks should slope away from the house.

A sidewalk is about ready for pouring. This is the one that
was four inches lower on the end next to the porch
(Step No. 42).

STEP NO. 43

TERMITE-PROOFING UNDER PORCH. When
the porches are ready for pouring, termite-proofing material is sprayed just as it was for the floor slab. Not all
builders spray under the porches, but it is a wise precaution.

Pouring the front porch (Step No. 44).

STEP NO. 44

POURING THE PATIO, PORCHES, WALKS, AND DRIVES should be done on a clear day. Rain started falling just as we finished pouring our driveway, damaging the surface and making it impossible to smooth it out completely.

Draining also is important, as in step No. 42, to make sure the water drains away from the house.

Labor on this third phase of concrete work (Steps No. 42 and 44) totaled $347.02. Ready-mix concrete and form lumber totaled $819.95.

STEP NO. 45

GRADING. When the drives, porches, and walks—and retaining walls, if any—are completed, a tractor subcontractor is brought in to grade the lot and smooth it out around the concrete. If earth is not pushed up around the concrete, erosion might wash out the sand from under the slab and cause it to break. This is not final grading, and it takes only a couple of hours or so. The cost was $20 for our house.

Grading around the concrete work (Step No. 45).

STEP NO. 46

CONCRETE SAWING. A large area of concrete will expand and contract with temperature changes, so it is necessary to saw the driveway into smaller sections to allow room for this and prevent cracking and breaking. The concrete subcontractor does this as part of Phase III concrete work. Rental of the concrete saw was $20.55, but there was no additional labor charge.

STEP NO. 47

FINAL PLUMBING (Phase III) in
lation of the garbage-disposal unit, the

Installing the hot v

Pouring the back porch (Step No. 44).

Another shot of the back porch (Step No. 44).

The driveway takes shape, just before it started to rain
(Step No. 44).

Installing the dishwasher (Step No. 47).

water tanks, a gas outlet in the fireplace for starting a wood fire, and the bathtub, shower, washbasins, and commodes. Specifications and brands for these various items are listed in the specification sheets that go with stock blueprints.

The dishwasher cost $316.21. The plumber provided all other materials and labor at a cost of $1,259.39 for this final phase of his work.

STEP NO. 48

THE KITCHEN STOVE will be installed by the plumber if it is a gas stove, and by the electrician if it is electric, as a part of the final step for either subcontractor. The exhaust fan over the stove, however, is usually installed by the trim carpenter (Step No. 49). This normally involves only sawing a hole the size of the vent and slipping the vent into place, so it could be done by any of the above subcontractors.

We didn't like the built-in stove and oven (separate units) in our previous house, so we installed a regular kitchen range 30 inches wide in a slot exactly that size. It looks like a built-in, but is a single unit rather than scattered out over half of the kitchen. The total cost was $294.40, not counting installation labor, which was part of Step No. 47.

The kitchen stove fits neatly into the counter, much like a built-in, but is to us more compact than the usual built-in (Step No. 48).

91

STEP NO. 49

INSTALL KITCHEN STOVE VENT. The vent, or exhaust fan, usually is installed separately from the stove itself by the trim carpenter, because it involves carpentry rather than plumbing or electrical work. The job requires sawing a hole the size of the vent in the bottom of the cabinet and in the ceiling above the stove, then slipping the metal vent through the hole. This will draw the kitchen odors and smoke into the attic. The labor cost is negligible and is figured as part of the trim carpentry.

STEP NO. 50

FINISHING ELECTRICAL, HEATING, AND COOLING FIXTURES includes the installation of light fixtures ordered weeks or months earlier, wall switches, and outlets. Most electrical supply houses allow builders a 50 percent discount on fixtures. The final cost in our case (after discount) was $477.92, including switches, outlets, and plates.

About the only thing to watch out for here is that the light switches are put in the right places. The subcontractor will follow the blueprint, but in some instances the location of wall switches needs to be changed. Also, we had the subcontractor put a television antenna wire in the corner where we knew the set would be placed, which was considerably neater than having the wire showing.

The cost for this half of the electrical, heating, and cooling work came to $1,427.50 for labor and materials supplied by the subcontractor. This was the same amount —and the same subcontractor—as in Steps No. 14 and 15.

Installing the dining-room light fixture (Step No. 50).

STEP NO. 51

INSTALL GLASS AND MIRRORS. Custom cutting and installation of glass and mirrors is cheaper than buying stock materials. For our house, a local firm furnished, cut, and installed windows in two china-cabinet doors, two pieces of frosted glass on each side of the front door, and four large and six small pieces of mirror for $245.45.

An electrical outlet is installed in the wall (Step No. 50).

STEP NO. 52

COMPLETE TRIM WORK AND HARDWARE INSTALLATION. The final phase of trim carpentry comes next, including the installation of doorknobs and locks, towel racks, drawer and cabinet knobs and locks, closet rods, and other hardware that goes on after the finish is applied to the woodwork. For all phases of trim work (Steps No. 36, 49, and 52) we paid 75¢ a square foot plus $170 for extras above the bid price, for a total of $2,225 for labor. The total cost for materials was $4,522.49.

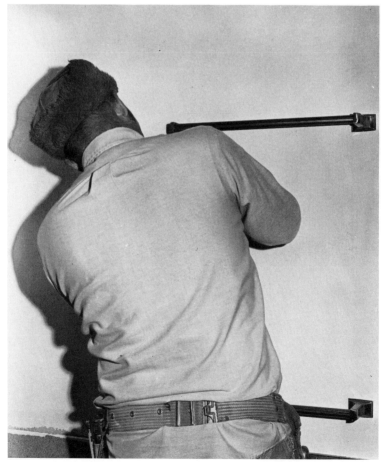

Installing towel racks (Step No. 52).

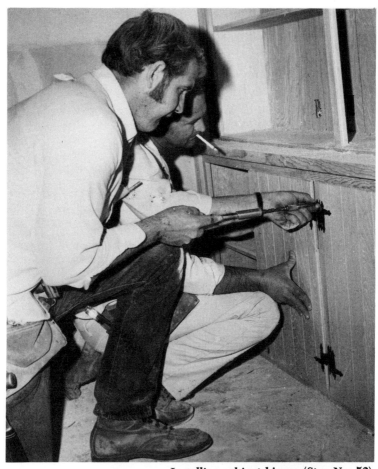

Installing cabinet hinges (Step No. 52).

A closet door lock goes into place (Step No. 52).

STEP NO. 53

CARPETING. Firms specializing in carpeting new homes offer a special rate to builders well below the prices charged by retail firms. Our shag carpeting and pads (194 square yards) came to only $5.10 a square yard, and the bathroom carpets and pads (21⅓ square yards) cost only $5.50 a square yard. Labor on installing the shag was 90¢ and for the baths $1.00 a yard. After adding $45 for repairs to the slab floor, all coverings (not counting brick in entry, hall, kitchen, and dining and utility rooms) totaled $1,380.85—for a quality well above that normally used by builders.

Splicing the living-room carpeting (Step No. 53).

Trimming a piece of carpet to fit a spot (Step No. 53).

STEP NO. 54

INSTALL TELEPHONE OUTLETS. At this point the telephone company finished the job of installing the telephone outlets. We ordered one fixed phone in the kitchen, and a second phone that can be moved to any one of four jacks.

STEP NO. 55

FINAL CLEANUP, inside, was done by a woman who specialized in that sort of thing. The job included cleaning and scraping the windows, vacuuming the carpet, washing brick floors, and cleaning the sawdust and other material from the cabinets and closets. So skilled was she that the job took only 22 hours at $3.25 an hour, totaling $71.50, the best bargain of the entire building project. When she was finished, the house was ready to move into.

STEP NO. 56

INSULATE THE ATTIC. The final step inside the house was the insulating of the attic, which was done by a giant motor blowing 6 inches of insulation material from a truck outside. This was done in the winter and immediately reduced the heating bill by 50 percent of what it had been the previous month. Materials and labor came to $210.

Feeding sacks of insulation material into a blower system that will transport it from this truck outside the house to the attic (Step No. 56).

Insulation arrives in the attic via this flexible tube. 6-inch
layer will cover the entire attic (Step No. 56).

STEP NO. 57

CONSTRUCT RETAINING WALLS.   Sometimes a
building site has so much slope that it is necessary to run
a retaining wall along one side. We had to build one
along 92 feet of our west property line, at a cost of $1 a
foot for labor (putting up concrete forms and inserting
steel reinforcing rods) and $22.39 for the concrete. The
wall ranged from a few inches to 2 feet high, so that only

one form was needed even at the highest point. If two
forms—up to 4 feet—had been needed, the cost per foot
for labor would have doubled.

Putting up the forms for a retaining wall (Step No. 57).

STEP NO. 58

FINAL GRADING includes leveling and grading the
entire yard with a tractor. The most important factor
here is contouring it so that the water will drain off the
yard and away from the house. Topsoil trucked in is
scattered out by the tractor without altering the contours.
For our house the tractor work cost $146 and the top
soil $300.20.

STEP NO. 59

DECORATIVE WALLS can be added as a final step,
around trees where the soil cannot be leveled out without damaging the trees, and elsewhere. We used a thou-

sand bricks at $69 and paid $377 for labor for two planters, a 15-foot circular wall around a part of a tree clump, and about 50 feet of wall around the back porch.

This picture of the back porch shows how the dirt was piled up at the near end before it was finally leveled out. The drainage pattern was center bottom of the photo to the center left, or along the porch about 3 feet out (Step No. 58).

The living room as seen from the entry foyer.

The completed fireplace.

The dining room as seen from the kitchen.

The master bedroom used as an office or den.

The dining room as seen from the living room.

The breakfast bar and kitchen as seen from the dining room.

chapter 8

Conclusion

I wouldn't want to mislead any prospective do-it-yourself home builder into thinking there are no problems involved in such an undertaking. We encountered quite a few, mostly small, some large.

Our biggest problem was in getting subcontractors who would do the job when they agreed to. Most subcontractors work regularly for a few contractors. Several agreed to work on our house, apparently just in case they might need the work when we needed them, and then backed out at the last minute when their regular contractors had jobs for them.

There is no way around this problem for the individual home builder. Because of this, it took us eight months to build the house a regular contractor would have completed in perhaps half that time. Patience, then, is vital, though at times one has a strong desire to kick someone's posterior.

Another problem, also with subcontractors, was the mistakes a few made. When the mistakes are spotted before the job is paid for, payment can be held up until corrections are made. This type of pressure isn't needed with reputable subcontractors, who will come back and make good just to maintain their professional reputations for quality work.

We spent about $600 extra correcting mistakes that some subcontractors left behind, and some poor work never did get corrected. We even had one subcontractor who vanished after completing his portion of the work, so we couldn't get him back to redo the sloppy work of his crew. We used a couple who blamed the mistakes on others and wouldn't make good. But most of our subcontractors either did quality work to start with, or came back willingly and without extra charge to make the desired improvements. Overall, our house came out better than most we have seen, with no major flaws.

The best safeguard is to hire only well-known and reputable subcontractors who have been in business a long time. In some instances, of course, workmen of this quality are not available, and the one-time builder will have to gamble on someone without a high reputation.

Another major problem comes from "extras" that subcontractors charge above their bid prices. We paid more than the bid price to about half of the subcontractors to cover items not in the blueprint on which they made their bids. Anything not on that blueprint will cost extra, and we really couldn't blame the subcontractor working on a fairly close margin. The only answer is to keep the extras as few and inexpensive as possible. In our case they came to about $1,000.

One word of caution: an amateur has no business trying to do skilled work himself. I tried at one point, and the results looked amateurish.

All things considered, we gambled and came out pretty well. We now have a home we couldn't have afforded any other way. I even enjoyed doing it, although it placed a bit of a strain on our marriage, as home building generally does. But I'd do it again, and may some day.

appendix 1
Subcontractors

1. Home designer (Step No. 2)

2. Tractor work (Steps No. 7, 41, 45, 58)

3. Sand and dirt hauling (Steps No. 16, 41, 44, 58)

4. Concrete work—footing and stemwall (Steps No. 8 to 12)

5. Concrete work—slab (Steps No. 16, 18, 20, 21)

6. Concrete work—porches, walks, driveways, and retaining walls (Steps No. 42, 44, 46, 57)

7. Plumbing—drains, water lines, washbasins, bathtubs, showers, commodes, and dishwasher (Steps No. 13, 15, 25, 47, 48)

8. Electrical, heating, and air conditioning (Steps No. 14, 17, 26, 50)

9. Termite spraying (Steps No. 19 and 43)

10. Rough-in carpentry (Step No. 23)

11. Insulation—walls and attic (Steps No. 29 and 56)

12. Roofer (Step No. 24)

13. Bricklayers (Step No. 34)

14. Trim carpenters (Steps No. 36, 49, 52)

15. Drywalls—sheetrocking, taping, plastering, and blowing texture material on ceilings (Steps No. 30 and 31)

16. Garage door and attic ladder installation (Step No. 32)

17. Finish work—painting and staining (Steps No. 31, 33, 37)

18. Cleanup (Steps No. 40 and 55)

19. Floor brick and tile, bathroom wall tile (Step No. 35)

20. Carpeting (Step No. 53)

21. Cabinet top installation (Step No. 38)

22. Vanity top installation (Step No. 39)

23. Glass and mirror installation (Step No. 51)

24. Telephone installation (Steps No. 27 and 54)

25. Decorative retaining walls (Step No. 59)

Summary of Steps in Home Building

(Note: Prices shown in parentheses are for items that were not
included in the $55,000 contractor's bid)

	Steps	Subcontractor Number	Labor	Materials	Other Labor and Materials Combined
Chapter 1 Precon- struction	No. 1 Home site selection			($3,490.00)	
	No. 2 House plans	1			($342.52)
	No. 3 Secure financing				
	No. 4 Secure building permit			$ 396.33	
	No. 5 Buy builder's risk insurance			$ 216.00	
	No. 6 Open charge accounts with lumberyards, builders' supply, and ready-mix concrete firms, etc.				
Chapter 2 The Base (floor slab and below)	No. 7 Site preparation	2	$ 45.00		
	No. 8 Stake out the house site	4	See No. 10		
	No. 9 Dig footing	4	See No. 10		
	No. 10 Put in steel rods	4	$ 197.31	$ 105.74	
	No. 11 Pour footing (concrete work, Phase I)	4	See No. 12	$ 570.66	
	No. 12 Set forms, pour stemwall	4	$ 242.25	$ 153.56	
	No. 13 Sewer lines below the floor slab (plumbing, Phase I)	7	See No. 15		
	No. 14 Electrical, heating, and air- conditioning groundwork—below the slab (electrical, Phase I)	8	See No. 26		
	No. 15 Water lines below the slab	7			$ 787.94
	No. 16 Fill sand below the slab	3 and 5	$ 246.00		
	No. 17 Install temporary electrical pole	8	See No. 26		
	No. 18 Build forms for various floor levels	5	See No. 21		
	No. 19 Termite proofing (Phase I)	9			$ 40.00
	No. 20 Place insulation around stemwall	5	See No. 21	$ 30.60	
	No. 21 Pour floor slab (concrete work, Phase II)	5	$ 431.16	$ 925.81	
Chapter 3 Walls and Roof	No. 22 Order brick, lights, windows, floor tile and brick, bath tile and other items that might take time to get				
	No. 23 (A) Rough-in carpentry	10	$ 2,119.40	$ 5,157.02	
	(B) Installation of windows	10		420.23	
	No. 24 Roofing	12	$ 412.00	$ 2,665.70	
	No. 25 "Top Out" plumbing (plumbing, Phase II)	7			$ 682.46
	No. 26 Electrical, heating, and air conditioning (electrical, Phase II)	8			$ 1,427.50
	No. 27 Pre-wire for telephone	24			$ 12.24
	No. 28 Electrical inspection				
	No. 29 Insulate walls	11			$ 208.00
	No. 30 Sheetrock, tape, and plaster	15	$ 717.64	$ 858.25	
	No. 31 "Blow" textured materials on the ceiling	15 or 17			$ 175.00
	No. 32 Install attic ladder and garage door with electronic opener	16			$ 415.86
	No. 33 Texturize walls and paint base coat on outside trim (Painting, Phase I)	17	See No. 37		
Chapter 4 Bricking	No. 34 Brick outside walls and fireplace	13	$ 1,280.00	$ 1,300.87	
Chapter 5 Interior Finishing	No. 35 Lay brick floor and tile	19			$ 1,515.00
	No. 36 Trim carpentry (Phase I)	14	See No. 52		
	No. 37 Finish woodwork, paint inside walls and outside trim (Painting, Phase II)	17			$ 1,645.20
	No. 38 Install cabinet tops	21			$ 313.63
	No. 39 Install vanity tops	22			$ 305.62

	Steps	Subcontractor Number	Labor	Materials	Other Labor and Materials Combined
	No. 40 Cleanup, inside and out	18	$ 125.00		
	No. 41 Grading for drives, walks and porches; sand hauling	3 and 2	$ 27.00	$ 166.74	
	No. 42 Build forms for driveways, walks, porches (concrete work, Phase III)	6	See No. 44	$ 76.35	
Chapter 6 Outside Finish Work	No. 43 Termite proofing under porches (Phase II)	9	See No. 19		
	No. 44 Construct patio, porches, walks, drives (concrete work, Phase III)	3 and 6	$ 347.02	$ 819.95	
	No. 45 Grading around drives, walks and patio	2	$ 20.00		
	No. 46 Concrete sawing (driveway)	6	See No. 44	$ 20.55	
	No. 47 Finish plumbing (Phase III)	7			
	Dishwasher	7		$ 316.21	$ 1,259.39
	No. 48 Install kitchen stove	7	See No. 47	($294.40)	
	No. 49 Install vent for kitchen stove	14	See No. 52	$ 62.65	
	No. 50 Finish electrical, heating and air conditioning (electrical, Phase III)	8	$ 1,427.50	$ 477.92	
	No. 51 Install glass in China cabinet doors, and mirrors in bathroom	23			$ 245.45
Chapter 7 Final Steps	No. 52 Complete trim work—hardware, door handles, etc.	14	$ 2,225.00	$ 4,522.49	
	No. 53 Install carpets	20			$ 1,380.85
	No. 54 Install telephone outlets	24	$ 63.62		
	No. 55 Final cleanup—woodwork, floors, windows, inside cabinets, etc.	18	$ 71.50		
	No. 56 Insulate attic	11			$ 210.00
	No. 57 Construct retaining wall	6	($92.00)	($22.39)	
	No. 58 Final grading, dirt hauling	3 and 2	$ 146.00	$ 300.20	
	No. 59 Construct decorative walls and planters	26	($377.00)	($69.00)	
	Builder's fee		$ 1,500.00		
	Miscellaneous		$ 245.50	$ 146.37	
	TOTALS		$11,800.90	$19,710.20	$10,624.14
	OVERALL TOTAL				$42,135.24

Glossary

BUILDER: The person in charge of the overall construction, hiring and coordinating the subcontractors who actually perform the various phases of the work.

CONTRACTOR: *See* Builder.

CONSTRUCTION LOAN: A loan to the contractor or owner to pay for materials and labor during construction.

DRYWALLS: Installation of sheetrock on walls and ceilings, and taping and plastering of the joints of the sheetrock.

DUCTWORK: A system of tubes placed under the floor slab through which air moves to various parts of the house. Another series of tubes are placed in the attic for returning air to the heating-cooling unit.

FINISH PAINTING AND DECORATING: Application of the final surface to the walls (paint) and to the woodwork (stain and lacquer).

FOOTING: Concrete poured into an open trench dug in the ground, usually 18 inches deep and 12 inches wide, forming the lower portion of the foundation to support the stemwall and building frame.

FORM: Wood frame into which concrete is poured to make a floor.

FOUNDATION: The footing and stemwall combined, upon which is constructed the building frame.

FRAMING: *See* Rough-in carpentry.

PLUMBING, GROUNDWORK: Installation of sewer, gas, and water lines below the floor slab.

PLUMBING, TOP-OUT: Installation of gas lines to outlets, water lines and drains above the floor slab to the various sinks, washbasins, bathtubs and showers, hot-water tanks, and faucets.

PLUMBING, FINISH: Installation of washbasins, hot-water tanks, sinks, bathtubs, shower stalls, and faucets.

PRE-WIRING: Installation of wiring for telephone outlets.

ROUGH-IN CARPENTRY: Building the framework and outside sheathing of the walls, installing windows, outside door sills, ceiling joists and rafters, and outside trim.

SLAB: The concrete floor of the house and garage.

STEMWALL: The concrete poured into forms placed on top of the footing. The walls (framework) set on top of this stemwall, while the slab is poured inside it.

STUDS: The 2-by-4-inch timbers that form the outside wooden frame and the inside partitions, to which partition board, sheetrock, and similar materials are nailed.

SUBCONTRACTORS: Persons responsible for different portions of home construction. They bid on their portion of the job for a set amount, then hire their own crews on an hourly basis to do the work.

TILES: Small flat pieces of clay baked in kilns, used to cover floors (particularly baths) and walls around tubs, showers, and washbasins.

TRIM (FINISH) CARPENTRY: Installation of woodwork, cabinet work, panelings, and window and door trim.

WAINSCOTING: Wood paneling.